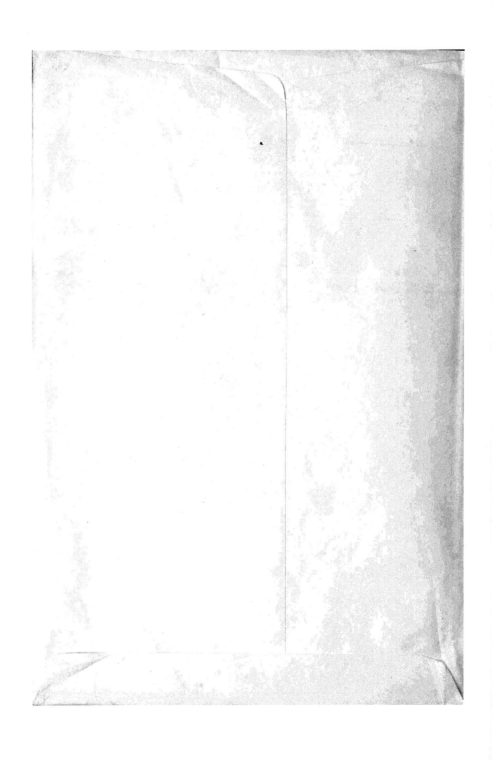

VICTORIA & ALBERT MUSEUM

DEPARTMENT OF TEXTILES

GUIDE TO
THE BAYEUX TAPESTRY

PRICE: ONE SHILLING NET.

HAROLD ENTHRONED. (*See* p. 6.)

Frontispiece]

VICTORIA & ALBERT MUSEUM.
DEPARTMENT OF TEXTILES

GUIDE TO
THE BAYEUX TAPESTRY

BY F. F. L. BIRRELL

LONDON : PRINTED UNDER THE AUTHORITY OF
HIS MAJESTY'S STATIONERY OFFICE, 1921

PREFATORY NOTE.

ALTHOUGH the Bayeux Tapestry is not itself included among the collections of the Museum, its surpassing interest to English students and the existence of two copies here amply justify the appearance of this guide. A full-size photographic reproduction, coloured by hand, is exhibited on the walls of the gallery numbered 79 on the first floor.

Another full-size photograph has been mounted on rollers for the convenience of students wishing to make detailed studies of the Tapestry; it may be seen on application at the Students' Room of the Department of Textiles, adjoining Room 123 on the first floor.

May, 1921.   CECIL H. SMITH.

5674

NOTE TO THE FIRST EDITION.

THE preparation of this guide has been entrusted to Mr. F. F. L. Birrell. Its chief aim is to present in a handy form the most important of the ascertained facts regarding the Tapestry. Matters which belong to the realm of conjecture, and even of controversy, have also been touched upon. Students wishing to pursue these further are recommended to consult the authorities quoted in the guide. I have read through the proofs, and must accept responsibility for such errors as may be found.

October, 1914. A. F. K.

NOTE TO THE SECOND EDITION.

WHEN this guide-book was first printed in 1914, its author, Mr. Francis Birrell, was for the time being a member of the staff of the Department of Textiles. On the publication of a second edition, they wish to express their regret that circumstances should have frustrated their hopes that he would remain as a colleague on the permanent staff.

April, 1921.

TABLE OF CONTENTS.

PAGE

THE BAYEUX TAPESTRY I

THE SCENES DESCRIBED 4

HISTORY OF THE TAPESTRY 11

THE LATIN INSCRIPTIONS WITH A TRANSLATION 14

THE PHOTOGRAPHIC COPY OF THE ORIGINAL 19

NOTES ON THE ILLUSTRATIONS 20

BIBLIOGRAPHY 28

INDEX 29

LIST OF ILLUSTRATIONS.

Frontispiece :—Harold Enthroned

PLATE   I.   King Edward the Confessor and Harold.

,,   II.   The Oath of Harold.

,,   III.   King Edward in his Palace.

,,   IV.   The Church of S. Peter at Westminster.

,,   V.   The Coronation of Harold. Stigand.

,,   VI.   The Comet.

,,   VII.   Building Ships.

,,   VIII.   A Feast.

,,   IX.   Odo. William. Robert.

,,   X.   Burning a House.

,,   XI.   The Battle of Hastings.

,,   XII.   The Death of Harold.

FIG. 1. A Lion   ..   ..   p. 1

,,   2. Ploughing   ..   ..   p. 9

,,   3. Harrowing   ..   ..   p. 14

,,   4 Fowling   ..   ..   p. 18

Fig. 1. A LION (*see* p. 27).

I.—THE BAYEUX TAPESTRY.

THE Bayeux Tapestry is probably the most famous and the most remarkable of mediæval embroideries. In it is given the complete story of a great exploit and one that must always be of particular interest to the English and French peoples—the story of the Norman Conquest. None of the main incidents leading up to the Conquest itself are omitted. The arrival of Harold in Normandy, his stay with William, his swearing of loyalty, his return to England, the death of Edward the Confessor, the subsequent election of Harold to the throne of England, with the expedition itself culminating in the Battle of Hastings, are all shown in the course of the story. It is as much a defence of William's conduct as a history of his triumph.

The Tapestry is generally considered to be contemporary, or almost contemporary, with the events it portrays, and it is evident that if such be the case it is an historical document of the very first importance. Certain archæologists have indeed maintained that it was not made till 50 or even 200 years after the Conquest, but their arguments have not generally been found convincing, and the claims of those who consider the Tapestry a contemporary document may be said to have held their ground. Historians like Freeman, Mr. J. H. Round, and Professor Oman have not hesitated to draw their

I

conclusions from it, and it remains an inexhaustible store-house of information regarding the social life of England and Normandy during the 11th century.

The interest of the Tapestry is still further increased when it is realised how much care seems to have been spent on giving an accurate rendering of the subjects shown. The representations of Edward and William, for instance, agree with their likenesses as shown to us on their seals and coins. In accordance with the fashion, Harold and his Saxons are given moustaches, while the Normans go clean-shaven. But while Harold and his men are in Normandy they follow the Norman fashion and also go clean-shaven. Again, the English " packed shield " formation, which so much impressed the Conqueror at Hastings, is shown with great clearness. Indeed, the Tapestry corrects many of the errors of mediæval historians, while Mr. Round believes that with its aid he has cleared up the mystery that hangs round early Norman " Castles." But though the Tapestry has these uses for archæologists and historians, it will have a wider appeal for those who are attracted by sound workmanship. Indeed, it combines beauty with an attention to correctness of detail and fashion to an extent unsurpassed in the memorials of the time.

A word may be added about technique. The " Bayeux Tapestry " always has been, and probably always will be, known under this name. It should, however, be recognised that, properly speaking, it is not a tapestry at all but an embroidery.* The figures are worked on linen, probably unbleached, in wool of eight different shades : dark and light blue, red, yellow, dark and light green, black and dove colour. The Latin inscriptions that are placed over most of the scenes are also stitched in wool, and are about an inch in height. It is noticeable that no attempt has been made to give the objects the colours they have in nature, while a difference of colour is the method employed for such effects as perspective.

* Exhibited with the photographic copy of the Bayeux Tapestry in Gallery 79 are two plaster reproductions made by Charles Stothard to show the technique of the embroidery. One piece shows the head of Duke William, and is taken from Scene 17. The second shows the head of Harold at his coronation, and is taken from Scene 33. In the Mediæval Department of the British Museum is a third cast, of the head of a soldier, which has been copied from Scene 23.

For instance, we see in the case of the horses the difference of the plane of their right and left legs is often suggested by a difference of colour.*

The tradition that the embroidery was carried out by Matilda and her ladies is a late tradition and need not be considered too seriously. William would have been more indisputably the hero if this were true, and it is not likely that the work, when completed, would have been sent to Bayeux to adorn Odo's cathedral. It is more probable that it was commanded by Odo for the decoration of his church. The design would be made by some competent artist in possession of the facts and this design would be worked out by the inhabitants of Bayeux and its neighbourhood. It should be remembered that it was very rare for the same persons to be concerned in the designing and the working out of the design.

* The Tapestry is 230 feet 9¼ inches long and 19⅝ inches broad. 1,512 objects of different kinds are shown in the course of the work, made up of 623 persons, 55 dogs, 202 horses or mules, 505 other animals, 87 buildings, 49 trees, and 41 ships and boats. The subject of each scene is given above in Latin.

II—THE SCENES DESCRIBED.

NOTE.—*In the following paragraphs the figures represent the number of the scene. Corresponding numbers will be found on the copy of the Tapestry above the scene described, and in Part IV. (p. 14).*

IT may be noticed, generally, that the story is presented with a strong bias in favour of the Normans, and there are large discrepancies between the account given here and that which, descending through Freeman from Florence of Worcester, is served up to English readers, whose sympathies are still supposed to be enlisted on behalf of Harold. It has been usually taught in England, for instance, that Harold only landed in France by mistake through being caught in a storm, and that this advantage was seized by William to the undoing of the man who was his guest. The story told in the Tapestry is very different.

In the first scene (PLATE I.) King Edward the Confessor is giving instructions to Harold, who immediately, with a large cavalcade of men, horses, dogs and falcons (2), sets off in the direction of the coast. He passes Bosham, where he enters the church and prays (3), conduct that is, perhaps, intended to appear hypocritical in the light of future events. After these religious exercises Harold and his company repair to a neighbouring house and refresh themselves with food before setting out again on their journey. They then cross the sea (4, 5, 6). No storm is shown and the Tapestry has the inscription " HIC HAROLD MARE NAVIGAVIT ET VELIS VENTO PLENIS VENIT IN TERRA WIDONIS COMITIS " (" Here Harold set sail upon the sea and with sails full of wind came to the land of Count Guy "). This cannot well mean a storm but a favourable voyage. Harold had always intended to come in to France. He had in the first scene had an interview with Edward in which he is presumably intrusted with a message to William, which he crosses the sea to deliver. If this message were to state Edward's wish that William should succeed

4

him on the Throne of England, how treacherous becomes Harold's conduct. Yet this is what the Tapestry silently suggests. On landing he is immediately seized (7) by Count Guy of Ponthieu, into whose territory he is come, and is carried off by him to his castle of Beaurain (8, 9).

Harold and Guy then converse together (10), Harold presumably explaining the reasons for his arrival in the latter's territories. The messengers of William arrive at Guy's castle (11), one of them, Turold, being mentioned by name. He is portrayed as a dwarf, either in an attempt at perspective or merely owing to the exigencies of space.* The story here becomes somewhat obscure, but the envoys go back to William (12, 13) and in 14, 15 Harold is brought to him by Guy in person. William's treatment of Harold is here surely intended to bring out the excellence of the Duke's character, for he behaves towards Harold with magnificent urbanity and takes him to his palace at Rouen (16, 17). In 18 appears the mysterious incident of " a certain clerk and Ælfgyva," which will be mentioned later. After this interval Harold and William, now apparently on the best of terms, start off together for a warlike expedition. They reach Mont Saint Michel (19), cross the River Couesnon (20), where Harold drags some Norman soldiers out of a quicksand (21) into which they had fallen. In 22 they arrive at Dol, and Conan† evacuates the town. They then pass Rechnes (Rennes), which is stitched small, and is only there to show the course of their march. A strenuous fight takes place at Dinan till Conan gives up the keys (23, 24), and in (25) William presents Harold with arms as a sign of gratitude for his services. In this high amity with each other they repair to Bayeux (26), where takes place the crucial incident in the history of Harold (27). Here there is again a great discrepancy between the Norman version and that usually taught in England. According to the English story Harold was tricked into swearing fealty to William on a box, in which, unknown to him, were concealed the bones of saints. This fact, if true, is suppressed in the Tapestry. Here, Harold's two hands are

* For Miss Agnes Strickland's theory that Turold was the designer of the Tapestry and Freeman's comments thereon, *see* Freeman's *Norman Conquest*, Vol. III., Appendix A. (2nd Edition).
† Conan II., Duke of Brittany, whose capital was Rennes.

resting on altars supporting chests of the kind used expressly for containing relics (PLATE II.). The oath is made as ostentatiously solemn as possible. Perhaps Harold might be able to urge compulsion as invalidating this oath; his whole position had been from the beginning little better than that of a prisoner. The plea of trickery was out of the question, and his future conduct makes him appear, to Norman eyes, a perjurer and a blasphemer.

After this Harold returns to England (28) and proceeds at once to Edward the Confessor (29), with whom he has an interview (PLATE III.).

We are next shown (30, 31) the coffin of Edward the Confessor being carried in mourning to its interment in the Church of St. Peter at Westminster (PLATE IV.). The order of events has been here slightly transposed, as in 32 we see Edward still alive though on his death-bed addressing his "fideles" or vassals. The reason for the transposition of scenes will soon become clear; Harold is present, with Edward's wife, Queen Eadgyth, an ecclesiastic, and two other persons. What may have been the nature of the communications passing between Edward and his "fideles" at this solemn moment cannot be said with certainty, but it is likely that even then the question of the succession was being agitated. Immediately below Edward has fallen into the sleep of death (ET HIC DEFUNCTUS EST), and the crown is offered to Harold, who sits enthroned, with Archbishop Stigand by his side (33) (PLATE V. and FRONTISPIECE). The reason why the funeral of the Confessor was represented before his death has now become apparent. The swift change from the death-bed of the saintly Edward to the triumph of his unscrupulous successor leaves behind it a feeling which must have been highly agreeable to Norman admirers of the Tapestry. In 34 (PLATE VI.) is recorded an interesting event. The inscription reads: "ISTI MIRANT STELLĀ" ("They," i.e. the English, "wonder at a star"), and this unusual star is portrayed curiously in the border. The English are right to be afraid. The heavens themselves blaze forth their disapproval at the conduct of Harold. A messenger is seen bringing tidings to the King, which he seems to hear with great agitation. It is likely that this message is connected with the strange apparition of the preceding scene. In 35 an

English ship is shown on its way to France, bearing to William news of events in England.

At this point the story as told in the Tapestry falls quite naturally into two parts. So far the unscrupulous conduct of Harold has been crowned with success. His ambitions are realised, and he sits on the throne of England. Omens, however, are not wanting to foreshadow the shortness of his reign. The second part of the story will show how William came into his own.

In 36 William has heard the news and promptly determines on revenge. With Bishop Odo at his side, he gives orders that a fleet shall be made ready ; the trees are hewn (37, 38), and the ships built (PLATE VII.) ; they are dragged down to the sea (39), arms are carried on board (40), carts with wine and arms are dragged down to the beach, William crosses in a great ship (41, 42, 43), he arrives at Pevensey (44, 45), the horses are disembarked (46), and the knights push on to Hastings (47). The whole story moves along with admirable speed, till the spectator seems to partake in the hurry and bustle of the great expedition. William's determination is as sure as Harold's. The difference is that his cause is just. After the arrival at Hastings, the story continues more slowly to its appointed end. That other side of military life is now shown, as necessary as, if less dignified than, the former glimpses. We see the victualling of William's army. In 47 the soldiers are seizing the neighbouring cattle for food. In 48 is a knight on horseback bearing the name Wadard. He is thought by the designer to be well enough known to need no explanation, but at this distance of time his appearance seems irrelevant. 49, however, carries on the story from 47. The food taken then is now being cooked, the servants serve up the meats which they lay before the soldiery, and Bishop Odo blesses the meat (50). This scene is worked in with real humour ; the soldiers are hardly able to restrain their appetite till grace has been said (PLATE VIII.). Immediately afterwards (51) are seen Bishop Odo, the Conqueror and his eldest son, Count Robert, the three most distinguished ornaments of the war, seated together in a tent (PLATE IX.).

In (52) it is ordered that a rampart be thrown up : the camp is shown. A messenger arrives giving news of Harold, and a house is burned (53). Then the army leaves Hastings

(54) to go in to battle against Harold (55, 56), and Duke William questions one Vital (57, 58) concerning Harold and his army. Not till 59 is news of William brought to Harold. That Harold should not have been kept more fully informed of his opponent's movements and only have heard particulars when the enemy were at his very gates seems to suggest a certain incompetence or, at any rate, to show that he had been completely surprised by the rapidity of William's movements. His other difficulties, caused by rivals to his throne and foreign enemies, which had only just been relieved by the glorious victory of Stamford Bridge, and the deaths of Tostig and Harold Hardrada, are omitted by the Norman chronicler. In 60 William exhorts his men to be brave and prudent, and they advance gaily into battle (61, 62, 63, 64, 65, 66), the whole course of events being splendidly exhibited. At last full justice is done to the English enemy. They fight valiantly, and the celebrated " packed-shield " formation is clearly shown (PLATE XI.). Bad luck, however, pursues the English ; Leofwyne and Gyrth, brothers of the King, are slain (67, 68, 69), but the fighting still remains even, Normans and English falling together (70, 71). The Normans are being pressed hard when Bishop Odo, armed with a club, exhorts the young recruits and saves a panic (72). A cry then goes up that Duke William has been killed ; he shows himself, however (73), and confidence is restored. On the left of the Duke is seen Eustace (E . . . TIUS is all that remains of the inscription) carrying a standard. According to the mediæval historian Benoît de Saint Maur* the Duke's standard-bearer Eustace, Count of Boulogne, had urged the Duke to leave the field, believing the day to be lost. This moment proves to be the turning point of the battle ; the French fight with renewed vigour (73, 74, 75), and beat their way up to Harold, who is killed by an arrow (76) (PLATE XII.). The English army is discouraged and flies (77, 78, 79).

* Benoît de St. Maur, the 12th century writer, is supposed to have been a native of St. Maur. From his prefix, Maître, he may have been a student at a University. But it is not known if he took orders. He was attached to the Court of Henry II. He was a loyal " Englishman," and always referred to the French as " they." He wrote the " Chronique des Ducs de Normandie," and also " Le Roman de Troie." The date of the " Chronique " is probably 1172–1176. The " Roman d'Enéis " and the " Roman de Thèbe " have also been attributed to him, but this is not generally accepted, while some even deny that the same hand composed the " Roman de Troie " and the " Chronique."

8

Here the Tapestry ends, and it is not quite clear if it had been intended to pursue the subject further. According to Dr. Ducarel the tapestry, when hung, exactly filled the nave of the Cathedral; so that very likely no more was ever designed.

The borders that run above and below well repay study. They not only make an admirable framework for the main narrative, but have an object of their own in keeping up a kind of running commentary on the events portrayed within their boundaries; strange birds and beasts, the subjects of fable, hybrids, and human forms, sometimes pursuing their ordinary avocations, sometimes engaged in battle or lying dead, form these borders, teaching by their actions, attitudes or expressions, the import of each scene. They express the hopes and fears of the rival factions and perform almost the functions of a Greek Chorus. They are delightful, too, in themselves, and there is in them something truly expressive of the mediæval mind.

Fig. 2. PLOUGHING (*see* p. 27).

There are several interesting features to be observed in connection with the Tapestry besides that of its evident bias. There runs throughout the assumption that the story will be familiar not only in outline, but also in detail to the examiners of the Tapestry—a fact which is in itself strong evidence of a contemporary date.

For instance, in 17 occurs the mysterious subject " UBI UNUS CLERICUS ET ÆLFGYVA " (" where a certain clerk and Ælfgyva "). Who Ælfgyva was permits of the widest conjecture; who a certain clerk may have been no one even pretends to know. But it is evident that the subject was sufficiently well known at the time to be inserted quite naturally and without any further explanation. At this distance of time it is impossible to explain the allusion. Again, who were Turold (12), Wadard (49), Vital (62) ? They are honoured in the Tapestry with their names above them, and so were evidently thought to be persons of importance. But few can have heard of them to-day. The archæologist Amyot, indeed, discovered that there were three vassals of Bishop Odo called by these names. If these are the people shown in the Tapestry, their appearance would be a compliment to the Bishop as well as themselves. In fact, throughout the story Bishop Odo appears with a prominence that can hardly have been attractive to his illustrious brother. Not only do his three servants appear in this way, but in 54 he is seated in state with the Conqueror and the Conqueror's eldest son, Count Robert, while in the crisis of the battle it was Odo, not William, who rallied the troops and turned into victory what had seemed certain defeat. Again, when William was giving his orders for the preparation of the Expedition (41) Odo stood by his side ever ready with advice. It may be remembered what William thought later of the ambition of his brother, and how some time after the Conquest was over he sent him packing back to his Bishopric. Odo was certainly a great benefactor to his Cathedral of Bayeux, and the prominence given to him has been used as an argument that the Tapestry was ordered by him and the design made by an artist intent on the gratification of his lord.

III.—HISTORY OF THE TAPESTRY.

THE " Bayeux Tapestry " has had an adventurous career since its first mention in the Inventory of Bayeux Cathedral in 1476, when it was hung round the nave during the season of the Feast of Relics. It is even possible that its adventures may have begun before this, if the assumption of an early date be correct, for the cathedral was burnt to the ground in 1106.

However that may be, in 1562 the town was sacked by Calvinists : but, fortunately, the Tapestry was handed over to the civil authority to guard, and it escaped destruction, though a tapestry " de grande valeur " that used to hang in the choir perished during the troubles. When these disturbances were over, it was once more in the hands of the ecclesiastical authorities, hung in the nave on appointed days, and forgotten for close on 200 years.

In the year 1724 an archæologist, M. Launcelot, read a paper before the French Academy on this subject. He had, however, only seen a drawing of a portion of the whole, and was only able to conjecture that the original was a fresco or an embroidery. He was strongly of the opinion that the original was made in the time of the Conqueror or his immediate successors. Better results, however, attended the efforts of Père Montfaucon, a Benedictine of St. Maur, who ran the original to earth after much search. It was published in engraving on a reduced scale in his second volume of "Monuments de la Monarchie Française (1730)." Kept in the repositories of the cathedral and only exhibited on feast days, the Tapestry survived in peace the early days of the Revolution, but when the Revolutionaries were going out to scatter their foreign enemies it was turned to account and made to cover an army waggon. It had been laid in position and was on the point of being taken off to the front, when M. le Forestier, the Commissioner of Police, seized on it and hid it in his study. In 1794 it was again about to be cut to pieces, when it was rescued by a self-appointed committee for guarding works of art in the neighbourhood.

The Tapestry was not unknown to Napoleon, and in 1803 it was sent to Paris and exhibited in the Musée Napoléon, doubtless with the intention of stirring the enthusiasm of the French into emulating the illustrious deeds portrayed. It was, however, returned to Bayeux in 1804 and deposited in the Library, with permission to be hung in the cathedral. fifteen days a year, a concession to the Church party that was never put into effect. It was exhibited in the Hôtel de Ville in 1830, and is now to be seen in a room built for it in the Public Library in 1842.

In 1871 on the near approach of the Prussians, the Tapestry was hastily taken down and hidden secretly away. When danger was passed it was returned to its former position. The Bayeux authorities, however, refuse to divulge the secret of its hiding-place, feeling that should adverse circumstances again arise it would be advisable that there should again be this secret spot in which to stow away the Tapestry.

The Bayeux Tapestry has since the 18th century received notice from English archæologists ; in 1746, Stukeley, author of the *Palæographica Britannica*, mentions it as " the noblest monument in the world, relating to our old English history." He was followed by a learned antiquarian, Dr. Ducarel, who stated that it was hung round the nave of the cathedral on St. John's Day, and continued there for eight days more. Two distinguished historians, Lord Lyttelton and David Hume, also discussed the Tapestry, the former being the first to doubt its contemporary date, thereby anticipating some modern criticism.

In the early years of the 19th century criticism of the Tapestry became more serious, the years 1816-1820 being very important in this respect. The views of Messrs. Stothard, Amyot, Hudson Gurney and others can be read in volumes XVIII. and XIX. of *Archæologia*.

In 1816 Mr. Charles Stothard was sent by the Society of Antiquaries to Bayeux to make a drawing of the Tapestry, and he brought home two small fragments with him.*   Within

* One of these fragments of the Tapestry was sold to Mr. Bowyer Nicholls and was purchased from him by the South Kensington Museum in 1864. It was soon decided to return this fragment to Bayeux, which was done in 1872. Mrs. Stothard has usually been accused of abstracting these two pieces. She was, however, able to show that she was not married to Mr. Stothard till 1818, the third and last year in which he visited Bayeux, and that at this date these fragments were already in his possession. Prior to his marriage he had possessed these two pieces, and said that they were so ragged as to be incapable of restoration. But how he had acquired them was not divulged.

two years he had completed his copy of the Tapestry, which is to be seen reproduced in Vol. VI. of the *Vetusta Monumenta*. Freeman, in Appendix A., Vol. III. (2nd edition), devotes a long passage to the subject and states his belief in its being made in England, an opinion which has not been generally shared.

Fig. 3. HARROWING (*see* p. 27).

IV.—THE INSCRIPTIONS.

THE Latin inscriptions above the embroidery run as follows (an English translation has been added, but no attempt has been made to amend or correct) :—

1.   EDWARD REX
   Edward the King.

2.   UBI HAROLD, DUX ANGLORUM, ET SUI MILITES EQUI-
         TANT AD BOSHAM .
   Where Harold, Duke of the English, and his soldiers
       ride to Bosham.

3.   ECCLESIA
   The Church (at Bosham).

4.   HIC HAROLD MARE NAVIGAVIT
   Here Harold crossed the sea.

5, 6.   ET VELIS VENTO PLENIS VENIT IN TERRA WIDONIS
         COMITIS
   And with sails full of wind came into the land of
       Count Guy.

6, 7.   HAROLD

7.   HIC APPREHENDIT WIDO HAROLDŪ
   Here Guy seizes Harold.

8, 9.   ET DUXIT EUM AD BELREM ET IBI EUM TENUIT
   And led him to Beaurain and held him there.

10.   UBI HAROLD (et) WIDO PARABOLANT
   Where Harold and Guy converse.

11.   UBI NUNTII WILLELMI DUÇIS VENERUNT AD WIDONĒ.
         TUROLD
   Where the messengers of Duke William came to
       Guy. Turold.

14

12. NUNTII WILLELMI
The messengers of William.

13. HIC VENIT NUNTIUS AD WILGELMUM DUCEM
Here the messenger came to Duke William.

14, 15. HIC WIDO ADDUXIT HAROLDUM AD WILGELMUM NOR-
MANNORUM DUCEM
Here Guy led Harold to William, Duke of the Normans.

16, 17. HIC DUX WILGELM CUM HAROLDO VENIT AD PALATIŪ
SUŪ
Here Duke William with Harold came to his Palace.

18. UBI UNUS CLERICUS ET ÆLFGYVA . . .
Where a certain clerk and Ælfgyva.

19. HIC WILLEM DUX ET EXERCITUS EJUS VENERUNT AD
MONTĒ MICHAELIS
Here Duke William and his army came to Mont
St. Michel.

20. HIC TRANSIERUNT FLUMEN COSNONIS
And here they crossed the river Couesnon.

21. ET HIC HAROLD DUX TRAHEBAT EOS DE ARENA
And here Duke Harold dragged them out of the
quicksand.

22. ET VENERUNT AD DOL ET CONAN FUGA VERTIT
And they came to Dol, and Conon turned in flight.

23, 24. REDNES. HIC MILITES WILLELMI DUCIS PUGNANT
CONTRA DINANTES ET CUNAN CLAVES PORREXIT
Rennes. Here the soldiers of Duke William fight
against the men of Dinan, and Conon reached
out the keys.

25. HIC WILLELM DEDIT HAROLDO ARMA
Here William gave Harold arms.

26. HIE (hic) WILLELM VENIT BAGIAS
Here William came to Bayeux.

27. UBI HAROLD SACRAMENTUM FECIT WILLELMO DUCI
Where Harold made an oath to Duke William.

28. HIC HAROLD DUX REVERSUS EST AD ANGLICAM TERRAM
Here Duke Harold returned to England.

29. ET VENIT AD EDWARDU REGEM
And came to King Edward.

15

30, 31. HIC PORTATUR CORPUS EADWARDI REGIS AD ECCLESIAM SCI PETRI APLI (Sancti Petri Apostoli)
Here the body of King Edward is borne to the Church of St. Peter the Apostle.

32. HIC EADWARDUS REX IN LECTO ALLOQUIT FIDELES
Here King Edward in bed addresses his vassals.
ET HIC DEFUNCTUS EST
And here he is dead.
HIC DEDERUNT HAROLDO CORONĀ REGIS
Here they gave to Harold the King's crown.

33. HIC RESIDET HAROLD REX ANGLORUM
Here sits Harold King of the English.
STIGANT ARCHIEP̂S (Archiepiscopus)
Archbishop Stigand.

34. ISTI MIRANT STELLĀ : HAROLD
These men are amazed at a star : Harold.

35. HIC NAVIS ANGLICA VENIT IN TERRAM WILLELMI DUCIS
Here an English ship came into the land of Duke William.

36, 37, HIC WILLELM DUX JUSSIT NAVES EDIFICARE
38. Here Duke William gave orders to build ships.

39. HIC TRAHUNT NAVES AD MARE
Here they draw down the ships to the sea.

40. ISTI PORTANT ARMAS AD NAVES
These men carry arms to the ships.
ET HIC TRAHUNT CARRUM CUM VINO ET ARMIS
And here they drag a cart with wine and arms.

41, 42, HIC WILLELM DUX IN MAGNO NAVIGIO MARE TRANSIVIT
43. Here Duke William crossed the sea in a great ship.

44, 45. ET VENIT AD PEVENESÆ
And came to Pevensey.

46. HIC EXEUNT CABALLI DE NAVIBUS
Here the horses go out of the ships.

47. ET HIC MILITES FESTINAVERUNT HESTINGA UT CIBUM RAPERENTUR
And here the soldiers hurried to Hastings to find food.

48. HIC EST WADARD
Here is Wadard.

16

49. HIC COQUITUR CARO
Here meat is cooked.
ET HIC MINISTRAVERUNT MINISTRI
And here the servants served.

50. HIC FECERUNT PRANDIUM
Here they made a feast.
ET HIC EPISCOPUS CIBŬ ET POTŬ BENEDICIT
And here the Bishop blesses the food and drink.

51. ODO EPŠ : WILLELM : ROTBERT
Bishop Odo : William : Robert.

52. ISTE JUSSIT UT FODERETUR CASTELLUM AT HESTENGA
The latter commanded that a rampart should be
    thrown up at Hastings.
CEASTRA
The Camp.

53. HIC NUNTIATUM EST WILLELMO DE HAROLD
Here news of Harold is brought to William.
HIC DOMUS INCENDITUR
Here a house is burned.

54. HIC MILITES EXIERUNT DE HESTENGA
Here the soldiers left Hastings

55, 56. ET VENERUNT AD PRELIUM CONTRA HAROLDUM REGE
And came into battle against King Harold.

57, 58. HIC WILLELM DUX INTERROGAT VITAL SI VIDISSET
    HAROLDI EXERCITŬ
Here Duke William asks Vital if he had seen Harold's
Army.

59 ISTE NUNTIAT HAROLDUM REGĒ DE EXERCITU WILLELMI
    DUCIS
This man informs Harold the King concerning the
Army of Duke William.

60, 61, HIC WILLELM DUX ALLOQUITUR SUIS MILITIBUS UT
62, 63, PREPARARENT SE VIRILITER ET SAPIENTER AD
64.   PRELIUM CONTRA ANGLORUM EXERCITŬ
Here William exhorts his soldiers to prepare themselves manfully and wisely for battle against the
English Army.

65, 66. The Battle.*
    * There is no inscription for these two scenes.

67, 68, HIC CECIDERUNT LEWINE ET GYRÐ, FRATRES HAROLDI
69.   REGIS
Here fell Leofwyne and Gyrth, brothers of Harold
the King.

70, 71. HIC CECIDERUNT SIMUL ANGLI ET FRANCI IN PRELIO
Here fell together English and French in battle.

72.   HIC ODO EP̄S BACULŪ TENENS, CONFORTAT PUEROS
Here Bishop Odo, holding a staff, rallies the young
troops.

73.   HIC EST WILELM̄ DUX
Here is Duke William.

73.   E . . . TIUS
Eustace.

73, 74, HIC FRANCI PUGNANT ET CECIDERUNT QUI ERANT
75.   CUM HAROLDO
Here the French fight and those who were with
Harold fell.

76, 77. HIC HAROLD REX INTERFECTUS EST
Here King Harold was slain.

78, 79. ET FUGA VERTERUNT ANGLI
And the English turned in flight.

Fig. 4.   FOWLING (*see* p. 27)

V.—THE PHOTOGRAPHIC COPY OF THE ORIGINAL.

ON the 3rd of August 1871 the Lords of the Committee of Council on Education agreed to Mr. Joseph Cundall going to Bayeux to obtain permission to take a full-sized photograph of the Tapestry. Permission having been obtained, a highly-skilled photographer, Mr. E. Dossetter, went to Bayeux for the purpose. In the first instance quite small photographs were taken, which were subsequently enlarged to the size of the original. A complete photographic copy enlarged to full-size and coloured after the original was exhibited in the Albert Hall at the Exhibition of 1873. This is the copy that is now exhibited in the Museum (Gallery 79).

What Carlyle thought of this copy cannot fail to be of interest—he expresses his enthusiasm in a letter to Sir Henry Cole :—

"I went yesterday with two companions for a look at your Bayeux Tapestry in the Albert Hall and I cannot but express to you at once my very great contentment with what I saw there. The enterprise was itself a solid, useful and creditable thing ; and the execution of it seems to me a perfect success far exceeding all the expectations I have entertained about it. Mr. Froude, who was one of my companions, was full of admiration, and a brother of mine who had seen the Tapestry itself at Bayeux last year seemed to think that this copy you had managed to make (I hope in a permanent and easily repeatable manner) was superior in vivid clearness, beauty of colour, etc., to the very original. As the work is in essence photographic, I flatter myself you have preserved the negative and other apparatus whereby the thing can be repeated as often as you like and at a moderate expense—in which case it might with evident and great advantage be imparted in the same complete form to all British Colonies, and even in America itself would be precious to every inquiring and every cultivated mind. In a word, I am much obliged to you for sending me to see this feat of yours (by far the reasonablest in completeness of its kind yet known to me), and very much obliged above all for your having done it and *so* done it.

"Yours truly, with many thanks,

"T. CARLYLE."

VI.—NOTES ON THE ILLUSTRATIONS.

PLATE I.

King Edward gives instruction to two persons, of whom one is Harold; Edward's clothes are richly embroidered. He is seated on a throne and has a crown and sceptre.

The scene shows a room in Edward's castle; a portion of the outside wall is given; but the rest is cut away to give a view of the interior in a manner very common in mediæval art. It will be seen that the castle is in the Norman style. On the left is a round Norman window and there are Norman turrets above. The throne on which Edward is sitting is typical of the art of the period, the animal's head which forms the right arm being a common decoration. In an Anglo-Saxon calendar of the 11th century (the MS. Cotton and Julius A VI.) a drinking party is shown on a large daïs, the two ends of which are in form like the head and front legs of two great dogs.

As to the nature of the communications passing between Edward the Confessor and Harold, it may be added that three reasons are given of Harold's journey to Normandy in different versions: (1) To release his brother and nephew from imprisonment; (2) that, owing to a storm when out fishing, he was shipwrecked on the coast of France; (3) to impart to William Edward's intention of making him his heir. The third was the Norman method of explaining what happened and is apparently the one accepted by the designer of the Tapestry.

PLATE II.

Harold's Oath.

William of Normandy, sword in hand, sits on the left while Harold takes a solemn oath of fealty; each hand rests on an altar, supporting a box of relics. The cloths that cover the altars are of embroidery, though in the Tapestry they look like velvet. The absence of any secrecy in the matter of the relics is here insisted on, the story being told from the Norman point of view.

The Palace of Westminster, and Edward the Confessor's body borne to the Church of St. Peter's.

Several points of interest arise in connection with these two plates ; it will be seen that the Palace of Westminster shows a general similarity to the representation in the first scene. The room in which Edward receives Harold is the same shape and similar turrets appear in each case. Further historical accuracy is shown in placing Edward's palace immediately to the left of St. Peter's Church.

This St. Peter's Church is the earliest form of what has since been known as Westminster Abbey, before it was rebuilt in the Gothic style during the latter part of the 13th century. This picture is of particular interest, as considerable care seems to have been taken in the reproduction to give an accurate picture. The long series of Norman arches below and the smaller row above are particularly noticeable. The centre tower is also well portrayed. The hand of God appearing through a cloud as if in dedication, and a man placing a weather-cock on the roof, seem to suggest that the church was just being completed, a fact that is nowadays believed to be true, though before it used to be held that Edward never finished the church.

The foundations of this original edifice are still to be seen in Westminster Abbey and in their main features bear a strong resemblance to the Abbey of Jumièges (see " Social England," Vol. I., p. 318), which was built about the same time and very likely designed by the same architect.

The whole question of the old building of St. Peter's at Westminster has been discussed by Professor Lethaby and the Dean of Wells, Dr. Armitage Robinson (formerly Dean of Westminster), in the Proceedings of the Society of Antiquaries for 1910.

PLATE V.

Harold enthroned* with Archbishop Stigand by his side.

Stigand wears an amice, a pall, a chasuble, gloves, boots, alb, stole and maniple. He is, however, bareheaded, which is an argument for an early date for the Tapestry as the mitre is rare in manuscripts till the 12th century. If the mitre had

* This portion of the plate also appears as the Frontispiece.

been known to the designer of the Tapestry he would surely
have given it to Stigand for so important an occasion. It will
be noticed that his chasuble is very long behind though very
short in front, and that his maniple is carried between his
thumb and first finger rather than hanging down from the
elbow, as is the fashion to-day. This fact goes to prove the
contention of those who hold that the maniple was originally
in the nature of a handkerchief. (The central portion of this
scene is reproduced on a larger scale in the frontispiece.)

PLATE VI.

"They wonder at a star," which is shown very curiously in
the border.

This appearance of Halley's Comet is mentioned by English,
Norman, South Gaulish, German and Italian chroniclers, by
whom it was generally held to portend the conquest of
England. It is also thought that there is a reference to it in
Chinese Annals. The comet appeared nine days after Easter,
and shone with great brightness for some days. Harold had
been crowned on the 6th of January, and the Conqueror
anchored off Pevensey on the 28th of September. Halley's
Comet also appeared in 1145, 1223, 1301, 1378, 1456, 1531,
1682, 1759, 1835 and 1910. A full account, with extracts
from the contemporary chroniclers, can be seen in Freeman's
"Norman Conquest" (2nd Edition). Vol. III., pp. 640–5.

PLATE VII.

Building the Ships.

The ships which are being built in this plate are the famous
"Snekkjur" or serpent vessels of the Vikings, so praised by
the Skalds. On the whole the Normans had altered their
marine equipment comparatively little since their Norwegian
days. A 9th century Viking boat was in 1880 discovered at
Gokstad on the west coast of the Gulf of Christiania. It was
about 75 ft. long, 16 ft. broad, 5·7 ft. deep, with a displace-
ment of 30 tons and able to carry 40 men. Such a ship
would have been a fine one in the days of the Conqueror.

William's fleet consisted of 3,000 boats of different shapes
and sizes, of which 696 were of the "Snekkjur" type. The
serpent decoration on the prow and stern of this type of boat

was often omitted, but is shown in the Tapestry. William's own ship was called the "Mora," and was a present to him from Queen Matilda. On the prow was the statue of a boy in copper gilt, who held a bow in his hand in which there was an arrow pointing ever towards England. His ship also flew the "Consecrated Banner" of Pope Alexander II., whose support of the expedition William had secured. ·

PLATE VIII.

A Feast is made.

On the left of this Plate chickens are being handed to the diners on spits, a spit apparently being provided for each guest. A knife is on the left table, also a piece of flat round bread, the common shape in which bread was made during that period. Spoons and forks were practically unknown at the time, and though two rough forks can be seen on a 12th century manuscript, the "Hortus deliciarum" of Herrad von Landsberg, they remain rare even in high society till the 16th century. Jean Sulpice writes as follows in 1480 on "La Civilité" :—" Prends la viande avec trois doigts et ne rempli pas la bouche de trop gros morceaux." A round bowl is on the table and one of the men drinks out of a large horn. These drinking horns were particularly popular in England, where practically no wine was drunk and mead was the favourite liquor. They were the same shape as musical horns with some form of stopper at the small end. Two horns of this kind are exhibited among the Ivories (Room 8) in this Museum (Nos. 7593, 8035-1862). A fine Rhenish drinking horn is in the British Museum, of which a good illustration is given in the article on Drinking Vessels in the " Encyclopædia Britannica." The British Museum also possesses a Scandinavian horn, illustrated on page 100 of the British Museum Guide to the Mediæval Department. The famous horn of Ulphus ("Homes of Other Days," p. 43) is in the Treasury of York Cathedral. Readers may remember in Boswell's " Tour to the Hebrides " how " we looked at Rorie More's horn, which is a large cow's horn, with the mouth of it ornamented with silver and curiously carved. It holds rather more than a bottle and a half. Every Laird of McLeod, it is said, must, as a proof of his manhood, drink it full of claret without laying it down." (Journal for Wednesday, Sept. 18th.)

These horns were not so universal in France, some forms of goblet being in use even during the 8th and 9th centuries, one such being seen on the right-hand end of the main table, where sit the Bishop and those of higher rank. By the 14th century horns seem to have gone completely out of use.

On this right-hand table is a goodly array of knives, goblets and plates of various sizes ; some pieces of the Norman pottery shown here are in the British Museum. The figure on the extreme right of this table, though occupying such an honoured place near the Bishop, is tearing a fish to pieces and thrusting it into his mouth with his fingers.

The figure coming towards the table holding a porringer is the cup-bearer and wine-taster, a prominent figure at every banquet. It is difficult to be certain as to the nature of the other object he is holding in his hand, but it may be a napkin, which was handed round to the most important people that they might wipe their fingers after the repast.

PLATE IX.

Odo. William. Robert.

This scene shows three portraits, William the Conqueror being in the centre, with Bishop Odo on his left and on his right Count Robert of Normandy.

On William's death, his son, William Rufus, succeeded to England, and Count Robert to Normandy. On the death of William Rufus in 1100, Count Robert was still on an expedition in the Holy Land. Hence Henry I. was elected king by the popular voice, in spite of protests from Normandy, and became an English as opposed to a French monarch.

This happy division of powers was, however, not permanent, and England, Normandy and many other French provinces were reunited under the Angevins, an event which marked the beginning of that perpetual trouble with France which hardly ended with King Henry VI.

PLATE X.

Burning a House.

In this plate again there seems to be a rough attempt at a portrayal of Norman domestic architecture. The Norman

24

nature of the work is suggested in the house being of two storeys, the Saxon dwelling usually being of one. A woman and her child escape from the hall, which occupies the whole of the ground floor, and is of nearly as much importance as it would have been in the Anglo-Saxon period. The room above is smaller and lit by a window with a Norman arch. No glass, however, would be placed in the window. The top storey would be provided with a rough kind of fire-place, as perhaps might also the ground floor, though the old fire piled right up in the middle of the hall was still quite common.

PLATE XI.

The Battle, showing the " packed shield " formation of the English.

This method of fighting particularly impressed William at Hastings, and no doubt the English employed it with great skill. But whereas the most important part of William's army was the cavalry, Harold's army consisted entirely of foot soldiers. The Thanes and other important men might be able to ride up to the scene of battle on horseback, but they dismounted for the fight. The " packed shield " formation they proceeded to employ consisted, as is seen in the Tapestry, of a thick wedge of men, widening out from about two in front to an uncertain number at the base ; the officers and better armed men formed the front wedge, backed by a dense column of the inferior troops.

The English and Normans wear for the most part the same armour, the body of which goes down to the knees in one piece, a type of armour known as the " hauberk " or " byrnie." These hauberks occasionally extended to the ankles, but the legs were generally cased in leather gaiters, somewhat resembling the " puttees " of to-day. Later, in the border, the hauberk is correctly shown being pulled off the the body of a dead soldier over the head, like a shirt.

In most cases the shields of the two armies are of the same shape, being pointed at the bottom and rounded at the top, a type that succeeded the kind which was narrow at each end and broadest in the middle. This old-fashioned shape of shield was still used by the English as well as the round embossed shield of yet earlier times.

An English warrior, probably Harold himself, is to be seen wielding one of those terrible battle-axes that did such execution at Hastings. This weapon was the mainstay both in attack and defence, and the glory of the Saxon army. But, nevertheless, it marks a far less advanced point in the history of war.

William's army is seen to be wearing stirrups, which, according to some authorities, were new to Europe at the beginning of the 12th century, having been introduced from China by the Mahommedans. But it seems extremely doubtful if they were such a late discovery as this. In any case, their use brought with it great changes in horsemanship, as the cavalry were enabled to sit forward on their saddles, often charging with their lances under their arm instead of leaning right back and charging with the arm erect.

It will be seen that the helmet is conical with a " nasal," that is to say, with a bar coming down as a shelter for the nose. The huge " vizor," covering the whole face and leaving only peep-holes for the eyes, was a later invention. Also the horses are here quite unprotected, not " tot couvert de fer " as Wace, a 12th century historian of the Conquest, would have had them be. The armour would certainly be very heavy, but lighter than the massive defences of the 14th and 15th centuries.

But the whole question of this early armour raises many points of difficulty and dispute. The subject of Saxon and Norman armour is well treated in Mr. C. H. Ashdown's " British and Foreign Arms and Armour " (1909), where special attention is devoted to the body-armour of the Bayeux Tapestry.

PLATE XII.

The Death of Harold.

The king is seen on the left, pierced through the right eye with an arrow which he is endeavouring to pull out with his hand. This is quite in accordance with tradition on the subject, though it is believed that Harold died, not at the moment of receiving the wound, but possibly some hours later. Three arrows are seen sticking in his shield. On the right there is an English warrior, struck down by the sword of a mounted Norman knight.

26

Figures *Nos.* 1 *to* 4.

These four figures form part of the borders under numbers X., XI., XII. ; three represent scenes from the ordinary life of the peasantry—ploughing, harrowing, and slinging at birds : the fourth shows the figure of a lion.

These rural scenes are extremely frequent in mediæval illustrated MSS. An illuminated MS. of the 11th century in the British Museum shows a good example of slinging (*see* " Social England," Vol. I., p. 316). The Louterell Psalter, which dates from the early years of the 14th century, gives many scenes of ploughing and harrowing (*see Vetusta Monumenta*, Vol. VI., especially Plates XXI. and XXII.). The shoulder collars and the iron shoes worn by the horses in these borders are held by some to have been an invention only made at the end of the 11th century.

VII.—BIBLIOGRAPHY.

Mr. F. R. Fowke has written a short and extremely clear account of the Tapestry in " The Bayeux Tapestry ; a History and Description " (George Bell & Sons, 1898), with a reproduction of the Tapestry. This invaluable book has been reprinted (1913) by Messrs. Bell in cheaper form. Professor Lethaby (" Embroidery, 1908–9 ") holds that the Tapestry may have been made in Kent.

In the study of the subject from an archæological point of view there are two admirable articles by Mr. J. Horace Round : " The Bayeux Tapestry " (" Monthly Review," December 1904) and " The Castles of the Conquest " (*Archæologia*, LVIII.). Mr. Round has also contributed an article on the Bayeux Tapestry to the " Encyclopædia Britannica."

In the *Archæological Journal*, Vol. LX., Sir W. H. St. John Hope has an article on " Fortresses of the 10th and 11th Centuries." All these books and articles support the belief that the Tapestry is practically contemporary with the events narrated.

M. le Commandant Lefebvre de Noëttes has approached the subject from a different point of view in the " Bulletin Monumental " of April 1912. He discusses the armour, weapons, harness of the horses and kindred subjects, coming to the conclusion that the Tapestry was probably made between 1120 and 1130, a date neither so late nor so early as extremists on either side have asserted.

The antiquity of the Tapestry has been attacked by M Marignan in his " La Tapisserie de Bayeux " (1902). He wishes to prove that its date cannot be before the middle of the 13th century. His views have, however, met with little support and have been answered by M. Lanore in his volume " La Tapisserie de Bayeux " (1903).

J. C. Bruce, in " The Bayeux Tapestry Elucidated with Coloured Illustrations " (1856), is inaccurate on some points but indulges in much ingenious speculation.

Two fresh books on the Tapestry have appeared in recent years. Mr. Hillaire Belloc (" The Book of Bayeux Tapestry," London, 1914), assigns the work to the second half of the 12th century. M. A. Levé (" La Tapisserie de la Reine Mathilde," Paris, 1919) holds that it was made for the consecration of Bayeux Cathedral in 1077.

INDEX.

A

	PAGE
Academy (French), paper before	11
Aelfgyva, personality unknown	5, 9
Albert Hall, Photograph of Tapestry at	19
Alexander II., Pope, blesses William's enterprise	23
Amyot the antiquary	10, 12
Angevins, Kings of England	24
Anglo-Saxon Calendar	20
Antiquaries, Society of, papers before	21
"Archæologia," papers in	12
Archæological Journal, papers in	28
Architecture, Anglo-Saxon	25
,, Gothic	21
,, Norman	20, 21, 24
Armitage-Robinson, Dr., Dean of Westminster	21
Ashdown, C. H., "British and Foreign Arms and Armour"	28

B

Battle-axes	26
Bayeux Cathedral, burned	11
,, ,, Inventory of	11
,, City, Hôtel de Ville at	12
,, ,, Public Library at	12
,, ,, sacked by Calvinists	11
Beaurain, Harold taken to	5
Benoît de Saint Maur, Annalist	8
Borders of Tapestry discussed	9
Bosham, Harold at Church of	4
Boswell, "Tour to the Hebrides"	23, 24
Boulogne, Eustace Count of	8
Boy on prow of William's ship	23
Bread	23
British Museum, Mediæval Horn at	23
,, Norman Pottery at	24

		PAGE
Brittany, Harold arrives in	.. ..	5
Bruce, J. C., Book on Bayeux Tapestry	.. ..	28
Bulletin Monumental, Article in	.. .. ..	28
Byrnie, Nature of	.. ..	25, 26

C

Calvinists sack Bayeux ..	.. ..	11
Carlyle, Thomas. Letter to Sir Henry Cole ..	..	19
Castles, Early Norman ..	.. ..	2
China, Annals of ..	.. ..	22
,, Spurs introduced from ..	.. .. ..	26
Christiania, Gulf of	.. ..	22
Chronique des ducs de Normandie. (Footnote)	..	8
Civilité, La, by Jean Sulpice ..	.. .. ..	23
Clerk, a certain, and Aelfgyva ..	.. .. ..	5, 9
Cole, Sir Henry. Letter from Thomas Carlyle	..	19
Comet, Halley's, appears	.. ..	6, 22
Conan II., Duke of Brittany ..	.. .. ..	5
Consecrated Banner, given by the Pope to William ..		23
Cousenon, R., Crossed by Harold and William	..	5
Cundall, Joseph, goes to Bayeux	.. .. ..	19

D

Dol, Siege of	.. ..	5
Dossetter, Mr., the Photographer	.. .. ..	19
Drinking Horns ..	.. ..	23, 24
Drinking Vessels ..	.. ..	24
Ducarel, Dr., Archæologist	.. ..	12

E

Eadgyth, Queen of Edward the Confessor	.. ..	6
Ecclesiastical Vestments worn by Stigand	.. ..	21, 22
Edward the Confessor ..	.. ..	1, 4, 6, 20, 21
Eustace, Count of Boulogne	.. ..	8

F

Feast of Relics, Tapestry exhibited during	.. ..	11
Florence of Worcester, Annalist	.. .. ..	4
Forestier M. le, Commissioner of Police in Bayeux ..		11

		PAGE
Forks		23
Fowke, F. R. Book on the Bayeux Tapestry ..		28
Fowling		25
Freeman, Professor J. H.	1, 4, 5, 13,	22
Froude, J. A., mentioned by Thomas Carlyle ..		19

G

Gaulish (South) Chronicles		22
German Chronicles		22
Gokstad, Viking ship discovered at		22
Gothic Style of Architecture		21
Gurney, Hudson, Antiquarian		12
Gyrth, Brother of Harold, killed		8
Guy Count of Ponthieu		4, 5

H

Halley's Comet		6, 22
Harold Hardrada defeated		8
Harold, King of England ..	1, 2, 4, 5, 6, 7, 8, 20, 21,	26
Harrowing		27
Hastings, Battle of	1, 8, 25,	26
,, Town of		7
Hauberks		25, 26
Henry I.		24
Henry II. (Footnote)		8
Henry VI...		25
Herrad von Landsberg		23
Holy Land, The		24
Hope, Sir W. H. St. John		28
" Hortus Deliciarum "		23
Hume, David		12

I

| Inscriptions on the Tapestry translated | 14, 15, 16, 17, 18 |

J

| Jumièges, Abbey of .. .. | 21 |

L

| Laird of Macleod, The, and Drinking Horns .. .. | 24 |
| Lanore, M., Book on Bayeux Tapestry .. .. | 28 |

Launcelot, M., reads paper before French Academy   11

Lefebvre de Noëttes, Commandant, Article on Bayeux

    Tapestry   28

Leofwyne, Brother of Harold, killed   8

Lethaby, Professor, Articles by..   ..   ..   .. 21, 28

Louterell Psalter, and Agricultural life   28

Lyttelton, Lord, on the Bayeux Tapestry   12

M

Macleod, Lairds of, and Drinking Horns   23

Mahommedans, spurs perhaps introduced by   ..   26

Marignan, M., book on Bayeux Tapestry   28

Matilda, Queen of William the Conqueror   23

Maur, Saint, Père Montfaucon of   11

Montfaucon, Père, Archæologist   11

" Monuments de la Monarchie Française " Publica-

    tions   11

" Mora," The, given by Matilda to William   23

Musée Napoléon, Bayeux Tapestry exhibited at   ..   12

N

Napoleon, and the Bayeux Tapestry   12

Nasal   26

Nicholls, Mr. Bowyer, Antiquarian   13

Norman Architecture 22, 23, 25

    ,,   Castles   2

    ,,   Chronicles   22

    ,,   Conquest 1, 8

    ,,   Pottery   24

O

Odo, Bishop of Bayeux 3, 7, 8, 10, 24

Oman, Professor   1

Omen at Harold's Coronation 7, 22

P

" Packed Shield " formation 2, 8, 25

" Palæographica Britannica," by Stukeley   12

	PAGE
Peter, Church of Saint, at Westminster	6, 21
Pevensey, William lands at	7
Ploughing	27
Ponthieu, Guy Count of..	4, 5
Prussians near Bayeux ..	12

R

Rennes, Harold and William pass	5
Robert, Count of Normandy	7, 24
Roman d'Enéis. (Footnote)	8
,, de Thèbe ,,	8
,, de Troie ,,	8
Rorie More's Horn	24
Round, Mr. J. Horace, Archæologist ..	1, 28
Rufus, William ..	24

S

Slinging	27
" Snekkjur " boats	22
Spoons	23
" Star," Appearance of strange	6, 22
Stigand, Archbishop	6
Storm, Question of Harold and, discussed	4
Stothard, Charles	8, 9
,, Mrs. Charles. (Footnote) ..	9
Strickland, Miss Agnes, on Bayeux Tapestry. (Footnote)	5
Stukeley. " Palæographica Britannica "	12
Sulpice, Jean, Author of " La Civilité "	23

T

Throne of King Edward the Confessor	20
Tostig, Death of	8
" Tour to the Hebrides," by James Boswell ..	23, 24
Turold	5, 10

U

| Ulphus, Horn of, in Treasury at York | 23 |

V

"Vetusta Monumenta," Reproductions in 13
Vital 8, 10
Vizor 26

W

Wace, 14th century Historian 26
Wadard 7, 10
Westminster, Church of Saint Peter at 6, 21
     ,,   Palace of 21
William the Conqueror 1, 3, 4, 5, 6, 7, 8, 9, 10, 20, 22, 24, 25 26

Y

York, Horn of Ulphus in Treasury at 23

PLATE I.

KING EDWARD THE CONFESSOR AND HAROLD. *(See p. 20.)*

PLATE II.

THE OATH OF HAROLD. (See p. 20.)

PLATE III.

KING EDWARD IN HIS PALACE. (*See* p. 21.)

PLATE IV.

THE CHURCH OF ST. PETER AT WESTMINSTER. *(See p. 21.)*

PLATE V.

THE CORONATION OF HAROLD. STIGAND. (See p. 21.)

PLATE VI.

THE COMET. *(See p. 22.)*

PLATE VII.

BUILDING SHIPS. (*See* p. 22.)

PLATE VIII.

A FEAST. *(See p. 23.)*

PLATE IX.

ODO. WILLIAM. ROBERT. (*See* p. 24.)

PLATE X.

BURNING A HOUSE. (*See* p. 24.)

PLATE XI.

THE BATTLE OF HASTINGS. (*See* p. 25.)

PLATE XII.

DEATH OF HAROLD. (*See* p. 26.)

Other publications of the Department of Textiles are shown below.

Publication No.

Catalogues.

72 T — English Ecclesiastical Embroideries of the XIII. to XVI. centuries. Second edition, with one illustration. pp. 45. Demy 8vo. 1911. 2d. [By post 3½d.]

117 T — Third edition, pp. viii and 47 ; 35 illustrations. Roy. 8vo. 1916. 9d. [By post 1s.]

115 T — Samplers. Second edition. pp. vii and 47 ; 12 plates. Roy. 8vo. 1915. 6d. [By post 8d.]

118 T — Algerian Embroideries. pp. 14 ; 4 plates. Roy. 8vo. 1915. 4d. [By post 5½d.]

91 T — Tapestries. By A. F. Kendrick. pp. 104 ; 19 plates. Crown 4to. Paper Boards. 1914. 1s. [By post 1s. 6d.]

129 T — Textiles from Burying Grounds in Egypt. By A. F. Kendrick. Vol. I, Graeco-Roman Period. pp. x and 142; 33 plates. Crown 4to. 1920. 5s. [By post 5s. 6d.] Vol. II. (*in the press*).

141 T — Franco-British Exhibition of Textiles. pp. 28 ; 18 plates. Crown 8vo. 1921. 9d. [By post 10½d.]

Guides.

96 T — Tapestries, Carpets and Furniture lent by the Earl of Dalkeith, March to May, 1914. pp. 27. Roy. 8vo. 1d. [By post 2½d.]

90 T — English Costumes presented by Messrs. Harrods, Ltd. pp. iv and 20 ; 16 plates. Roy. 8vo. 1913. 6d. [By post 8d.]

111 T — The Collection of Carpets. pp. viii and 88 ; 49 plates. Roy. 8vo. 1920. 2s. 6d. [By post 2s. 10d.] Cloth 3s. 6d. [By post 3s. 11d.]

119 T — Japanese Textiles. Part I.—Textile Fabrics. pp. xi and 68 ; 25 plates. Roy. 8vo. 1919. 3s. 6d. [By post 3s. 9d.]

120 T — Japanese Textiles. Part II.—Costume. pp. 65 ; 7 plates. 30 figs. Roy. 8vo. 1920. 3s. 6d. [By post 3s. 9d.]

136 T — Notes on Carpet-Knotting and Weaving. pp. 26 ; 12 plates. Crown 8vo. 1920. 9d. [By post 11d.]

Portfolios.

83 T — Tapestries. Part I. 1913. 6d. [By post 8½d.] Part II. 1914. 6d. [By post 8½d.] Part III. 1916. 1s. 6d. [By post 1s. 9½d.] In paper wrappers, 15 × 12. Each plate has descriptive letterpress on the attached flysheet.

2 Coloured Reproductions of English Silk Embroidery of the early 18th Century, 1913. 1s. each plate. 15 × 12. [By post 1s. 2½d.]

Lightning Source UK Ltd.
Milton Keynes UK
UKHW022319080223
416651UK00001B/37